Nurse Tank Failure With Release of Hazardous Materials Near Calamus, Iowa April 15, 2003

Hazardous Materials Accident Report

NTSB/HZM-04/01

PB2004-917001
Notation 7564A

National Transportation Safety Board

Washington, D.C.

this page intentionally left blank

Hazardous Materials Accident Report

Nurse Tank Failure With Release of Hazardous Materials Near Calamus, Iowa April 15, 2003

NTSB/HZM-04/01
PB2004-917001
Notation 7564A
Adopted June 22, 2004

National Transportation Safety Board
490 L'Enfant Plaza, S.W.
Washington, D.C. 20594

National Transportation Safety Board. 2004. *Nurse Tank Failure With Release of Hazardous Materials Near Calamus, Iowa, April 15, 2003.* Hazardous Materials Accident Report NTSB/HZM-04/01. Washington, DC.

Abstract: About 11:50 a.m. central daylight time on April 15, 2003, a nonspecification cargo tank used by River Valley Cooperative (River Valley) exclusively for agricultural purposes as a nurse tank split open after being filled with anhydrous ammonia at River Valley's nurse tank filling facility near Calamus, Iowa. About 1,300 gallons of the poisonous and corrosive gas escaped, seriously injuring two nurse tank loaders, one of whom died from his injuries 9 days after the accident. Equipment repair and replacement costs associated with the accident totaled about $3,100.

As a result of its investigation of the accident, the National Transportation Safety Board identified the following major safety issues: the adequacy of standards for initial qualification and periodic testing of nurse tanks, and the adequacy of River Valley's emergency procedures for anhydrous ammonia nurse tank loaders.

As a result of its investigation of this accident, the Safety Board makes safety recommendations to the Research and Special Programs Administration and River Valley.

The National Transportation Safety Board is an independent Federal agency dedicated to promoting aviation, railroad, highway, marine, pipeline, and hazardous materials safety. Established in 1967, the agency is mandated by Congress through the Independent Safety Board Act of 1974 to investigate transportation accidents, determine the probable causes of the accidents, issue safety recommendations, study transportation safety issues, and evaluate the safety effectiveness of government agencies involved in transportation. The Safety Board makes public its actions and decisions through accident reports, safety studies, special investigation reports, safety recommendations, and statistical reviews.

Recent publications are available in their entirety on the Web at <http://www.ntsb.gov>. Other information about available publications also may be obtained from the Web site or by contacting:

National Transportation Safety Board
Public Inquiries Section, RE-51
490 L'Enfant Plaza, S.W.
Washington, D.C. 20594
(800) 877-6799 or (202) 314-6551

Safety Board publications may be purchased, by individual copy or by subscription, from the National Technical Information Service. To purchase this publication, order report number **PB2004-917001** from:

National Technical Information Service
5285 Port Royal Road
Springfield, Virginia 22161
(800) 553-6847 or (703) 605-6000

The Independent Safety Board Act, as codified at 49 U.S.C. Section 1154(b), precludes the admission into evidence or use of Board reports related to an incident or accident in a civil action for damages resulting from a matter mentioned in the report.

Contents

Executive Summary

About 11:50 a.m. central daylight time on April 15, 2003, a nonspecification cargo tank used by River Valley Cooperative (River Valley) exclusively for agricultural purposes as a nurse tank split open after being filled with anhydrous ammonia at River Valley's nurse tank filling facility near Calamus, Iowa. About 1,300 gallons of the poisonous and corrosive gas escaped, seriously injuring two nurse tank loaders, one of whom died from his injuries 9 days after the accident. Equipment repair and replacement costs associated with the accident totaled about $3,100.

The National Transportation Safety Board determines that the probable cause of the sudden failure of the nurse tank at the anhydrous ammonia filling facility near Calamus, Iowa, on April 15, 2003, was inadequate welding and insufficient radiographic inspection during the tank's manufacture and lack of periodic testing during its service life.

As a result of its investigation of the accident, the Safety Board identified two major safety issues:

- The adequacy of standards for initial qualification and periodic testing of nurse tanks.

- The adequacy of River Valley's emergency procedures for anhydrous ammonia nurse tank loaders.

As a result of its investigation of this accident, the Safety Board makes safety recommendations to the Research and Special Programs Administration and River Valley.

Factual Information

Accident Synopsis

About 11:50 a.m. central daylight time[1] on April 15, 2003, a nonspecification[2] cargo tank used by River Valley Cooperative (River Valley) exclusively for agricultural purposes as a nurse tank split open after being filled with anhydrous ammonia at River Valley's nurse tank filling facility near Calamus, Iowa. (See figures 1 and 2.) About 1,300 gallons of the poisonous and corrosive gas escaped, seriously injuring two nurse tank loaders, one of whom died from his injuries 9 days after the accident. Equipment repair and replacement costs associated with the accident totaled about $3,100.

Figure 1. Accident nurse tank with shell fracture area circled.

---

[1] Unless otherwise specified, the times used in this report are central daylight time.

[2] The U.S. Department of Transportation publishes manufacturing specifications to which most tanks to be used to transport hazardous materials must be built. However, tanks to be used to transport hazardous materials for some specified activities, such as certain agricultural applications, do not have to meet these specifications.

Figure 2. Close-up view of accident nurse tank shell fracture area.

The Accident

The two nurse tank loaders involved in this accident were working for River Valley, a farmer's cooperative in eastern Iowa. They loaded anhydrous ammonia, a poisonous and corrosive liquefied gas, from a storage tank at the cooperative's Calamus filling facility into nurse tanks, which are nonspecification cargo tanks used for agricultural purposes. Once filled, each nurse tank was taken to a cooperative member's farm, where the anhydrous ammonia was injected into the soil.[3] Spring is the primary season for injecting anhydrous ammonia into farmers' fields. During the peak period[4] for this facility, about 150 nurse tanks per week were filled at the Calamus site. When this accident took place, the Calamus filling facility was in its peak activity period.

On April 15, 2003, the two nurse tank loaders, who were the only people at the facility at the time of the accident, had been filling nurse tanks from the storage tank, which was their usual task. In the late morning, they finished filling the accident nurse tank and disconnected the loading hoses and fittings. One loader was standing on the

---

[3] Farmers inject anhydrous ammonia into their fields before planting crops to increase the soil's nitrogen content. A nurse tank is typically pulled to the field using a pickup truck. In the field, the tank is attached to an anhydrous ammonia injector. The injector and nurse tank are pulled through the field by a farm tractor.

[4] The peak period is a 7- to 10-day span each spring.

loading platform, while the other had backed a pickup truck up to the full nurse tank and was standing at the back of the pickup truck hooking the nurse tank to the truck. Suddenly, the nurse tank shell split open at the bottom of its front half. The pressure of the escaping gas made several holes in the gravel lot surrounding the platform; the largest was about 7 feet long, 5 feet across, and 30 inches deep.

The force of the event threw the loader who had been standing by the pickup truck against the back of the truck. He could recall nothing after this until he woke to find emergency responders treating him.

The loader who had been on the loading platform could not tell investigators what occurred immediately following the sudden release of anhydrous ammonia because the injuries he sustained during the accident were so severe that he could not be interviewed by investigators before he died 9 days after the accident. This loader did tell a witness that he had activated the emergency shutoff button for the anhydrous ammonia pump to the other three nurse tanks on the platform. (See figure 3 for layout of Calamus filling facility.)

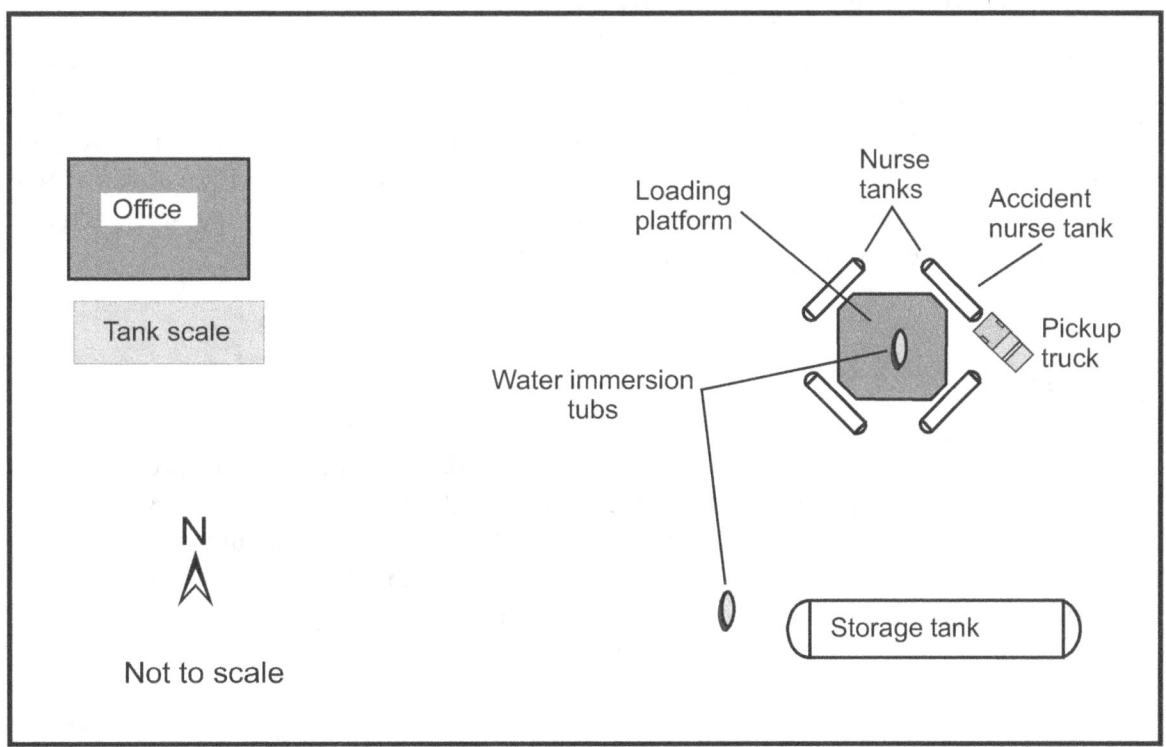

Figure 3. Layout of Calamus anhydrous ammonia filling facility.

Emergency Response

At 11:50 a.m., a woman who used this filling facility to provide anhydrous ammonia for her farm was driving by the Calamus facility when she saw a large, white vapor cloud emanating from the loading platform area. She pulled into the facility. She saw one tank loader lying in the emergency water immersion tub[5] on the platform and a second loader standing outside the tub assisting the one in the water.[6] She went into the facility's office, called 911, and told the 911 operator that an accident had occurred. She told the operator that anhydrous ammonia might still be leaking and that two people had suffered injuries.

Over the next few minutes, several people arrived at the facility and, finding the emergency situation, attempted to assist. One witness who arrived during this period recalled seeing vapor coming from under the loading platform. No other witnesses indicated to investigators that they had seen vapor in the vicinity at this time. With these people on the scene, the second loader got into the immersion tub with the first.

Fire department equipment and an ambulance were en route to the scene by noon and were on scene at 12:02 p.m. (By this time, the visible vapor cloud of anhydrous ammonia vapor had largely dispersed from the platform area.) After arriving on the scene, fire department representatives requested that a second ambulance and a helicopter be dispatched. The injured loaders were decontaminated[7] and stabilized on scene. At 12:21 p.m., a second helicopter was requested.

The first helicopter arrived at 12:45 p.m. and departed at 1:00 p.m. with the loader who had been in the immersion tub when the first passerby came on the scene. This loader was first taken to Davenport Hospital and then transferred to the burn unit of University Hospital in Iowa City, Iowa. The second helicopter landed after the first took off and picked up the loader who had been standing beside the tub when the first passerby arrived. This loader was taken directly to the University Hospital burn unit.

Injuries

Both nurse tank loaders suffered more than 50 percent body surface area chemical burns, eye injuries, and inhalation injuries due to anhydrous ammonia exposure. The loader who had been on the loading platform when the tank fracture occurred and who had been assisting his colleague from outside the immersion tub when witnesses arrived died as a result of inhalation injuries 9 days after the accident.

---

[5] A *water immersion tub* is used to provide on-site emergency first aid for people who experience injuries to the skin or eyes caused by exposure to corrosive materials, such as anhydrous ammonia. The tub is used to immerse in water the part of the body exposed to the corrosive material to reduce the tissue damage caused by the exposure.

[6] The loader in the immersion tub ultimately survived the accident; the loader standing outside the tub died as a result of his injuries.

[7] In this case, decontamination would have consisted primarily of showering the affected person with clean water and removing affected clothing.

Damage

As a result of the accident, the front half of the bottom of the nurse tank shell split open. The 53.5-inch-long split was located about 6 inches to the right of center at the tank bottom (viewed from the back). The split was along one side of a longitudinal weld in the shell on the tank bottom. (See figure 4.) Equipment repair and replacement costs associated with the accident totaled about $3,100.

Figure 4. Fracture running beside, to the left of, longitudinal weld (line of weld is marked with arrows).

Meteorological Information

The weather at the time of the accident was dry, the skies were clear, and the temperature was about 81° F. Winds were gusting from the southwest about 19 to 24 mph.

Nurse Tank Information

The accident nurse tank, serial number 1019945, was a non-U.S. Department of Transportation (DOT)-specification cargo tank built in 1976 to the requirements of the American Society of Mechanical Engineers (ASME) *Boiler and Pressure Vessel Code*: Section VIII, "Rules for Construction of Pressure Vessels," by Trinity Industries, Inc., (Trinity) at Beardstown, Illinois,[8] for use as an anhydrous ammonia nurse tank. Sometime after its manufacture, the tank was acquired by S/M Service Company. River Valley assumed ownership of the nurse tank during a 1999 merger with S/M Service Company. Until April 11, 2003, the nurse tank had been stationed at River Valley's Eldridge, Iowa, filling location. On that day, a farmer who had been using the tank in his field after obtaining it from the Eldridge facility dropped the tank off at the Calamus filling facility.

The nurse tank was made of SA-455 steel, which has a minimum specified tensile strength of 75,000 pounds per square inch (psi). The nominal thickness of the shell was 0.375 inch, and the nominal thickness of the heads was 0.362 inch. The tank was about 47 inches in diameter and about 205 inches (about 17 feet) long. It had a water capacity of 1,415 gallons. The heads were hot-formed and ellipsoidal in shape (concave to pressure). The maximum allowable working pressure of the nurse tank was 250 psi, gauge (psig) at 125° F.

The nurse tank was equipped with a 3/4-inch-diameter pressure relief device manufactured by Continental NH_3 Products, Inc., Dallas, Texas. This device was a spring-loaded valve designed to open at 250 psig +10 percent (between 250 and 275 psig). The nurse tank was also equipped with a float gauge manufactured by Rochester Gauges, Inc., Dallas, Texas, and a dip tube.[9] There were no openings on the nurse tank larger than 2 inches in diameter.

Tank shells are formed from two pieces of plate steel. Each plate is rolled into a cylinder or tube, until the opposite edges of the plate meet. The location where the edges meet is called a longitudinal seam. Each longitudinal seam is closed by welding both inside and outside the shell. These two cylinders are welded together with a circumferential weld to make a single cylinder, and the ellipsoidal steel heads are welded on each end of the single cylinder to form the nurse tank. Before closing, baffles are welded inside the tank to limit liquid surge during transportation.

The ASME *Boiler and Pressure Vessel Code* (to which the accident nurse tank was built) permits the use of spot radiography to check the welds for all pressure vessels constructed to the code and not containing "lethal substances."[10] Spot radiography consists of performing a 6-inch-long radiograph for every 50 feet of welds. When manufactured, the accident nurse tank's longitudinal welds were subject to spot radiography, in

---

[8] The Beardstown plant closed about 1986.

[9] A fixed-length *dip tube* is a small diameter pipe that extends from the top of the tank into the tank to the tank's maximum permitted loading level. The tube is equipped with a valve at or near the tank jacket. Loaders use the device to indicate when the liquid in the tank has reached the maximum permitted level.

[10] ASME defines *lethal substances* as "poisonous gases or liquids of such a nature that a very small amount of the gas or of the vapor of the liquid mixed or unmixed with air is dangerous to life when inhaled."

accordance with the ASME code's accepted spot radiography procedures. (See the "Regulations and Standards Affecting Nurse Tanks" section of this report for additional information about the ASME code and spot radiography.)

After manufacture, the accident nurse tank was hydrostatically proof pressure-tested to 375 psig.

Metallurgical Examination of the Nurse Tank

The metallurgical examination of the tank indicated that, when manufactured, a portion of the nurse tank's interior longitudinal weld was not centered on the shell seam. Examination of the nurse tank's 53.5-inch-long fracture, which was along the longitudinal weld, revealed four significant regions. (See figure 5.) They are described below:

- *Unfused region* - Where the inner weld bead[11] was offset to one side of the shell seam, an unfused region 3.25 inches long and up to 0.102 inch deep was found. (See figure 6.)

- *Oxidized region* - Beneath and encompassing the unfused region, the fracture surface contained a 12.75-inch-long region covered with black oxides, consistent with oxidation in a low-oxygen environment over a considerable period. This oxidized region had a smoothly curving boundary, consistent with the fracture's initiating from the vicinity of the unfused region. The oxidized region's maximum depth was 0.317 inch (including the 0.007-inch-wide "oxidized fatigue region" described below), and it came to within 0.063 inch of the tank's exterior surface. Cleaning with a deoxidizer revealed that this region (except the outer edge "oxidized fatigue region" described below) had relatively rough fracture features consistent with overstress or low-cycle fatigue fracture.[12]

- *Oxidized fatigue region* - Cleaning also revealed a 0.007-inch-wide region at the outer edge (in the direction of the tank exterior) of the oxidized region that had features consistent with high-cycle fatigue cracking, which is caused by repeated low-load events over time.

- *Overstress region* - The remainder of the fracture surface (to the breach in the exterior tank wall) had features consistent with overstress failure, such as failure caused by a single load event exceeding the load-bearing capacity of the cracked tank wall.

---

[11] The weld bead in the area of the unfused region was observed to make a large angle with the shell wall (as shown in figure 6). Such a feature is known to significantly increase the local operating stresses in a structure. See John F. Harvey, *Theory and Design of Modern Pressure Vessels*, Second Edition (New York, NY: Van Nostrand Reinhold Company, 1974) 348–349.

[12] Both overstress and low-cycle fatigue generate rough fracture features on a crack surface, as opposed to the smooth features generated by high-cycle fatigue. A tank that has experienced low-cycle fatigue has been exposed to infrequent high-load events, such as overpressurization or accident damage, rather than the single high-load event of an overstress fracture. Over time, corrosion may reduce or eliminate the evidence of the occurrence of multiple events, leaving what appears to be evidence of an overstress fracture.

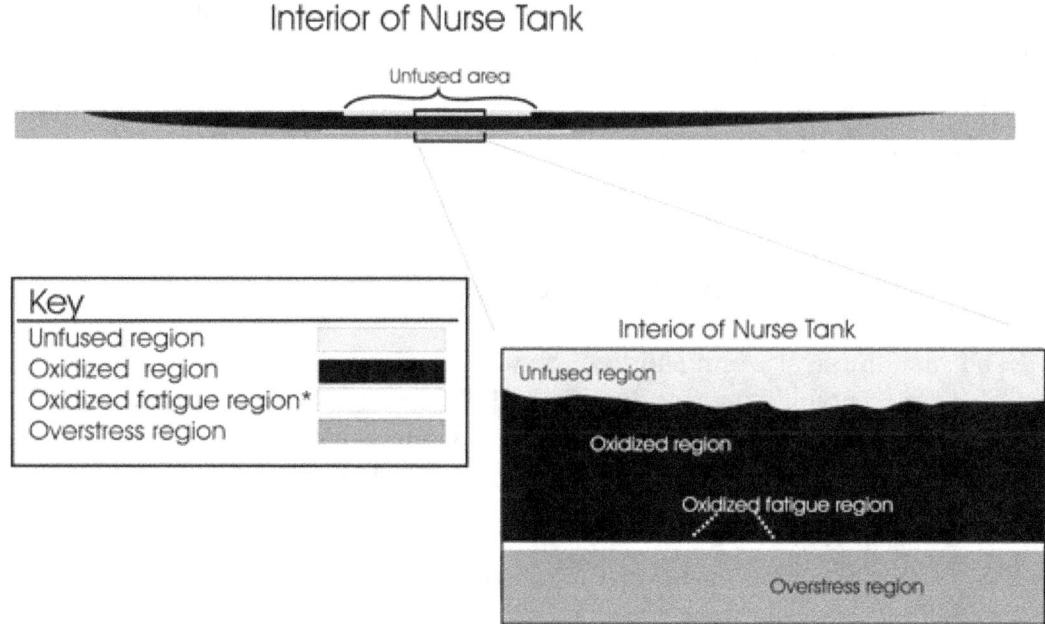

Figure 5. Diagram showing the four regions of the fracture surface. This diagram is not to scale and is intended only for use as a visual aid to the reader. [*Note: The oxidized fatigue region indicated above was actually part of the total oxidized region found during the postaccident examination. Because the fatigue region was metallurgically different from the rest of the oxidized region, it is indicated on this graphic by a different shade from the rest of the oxidized region.]

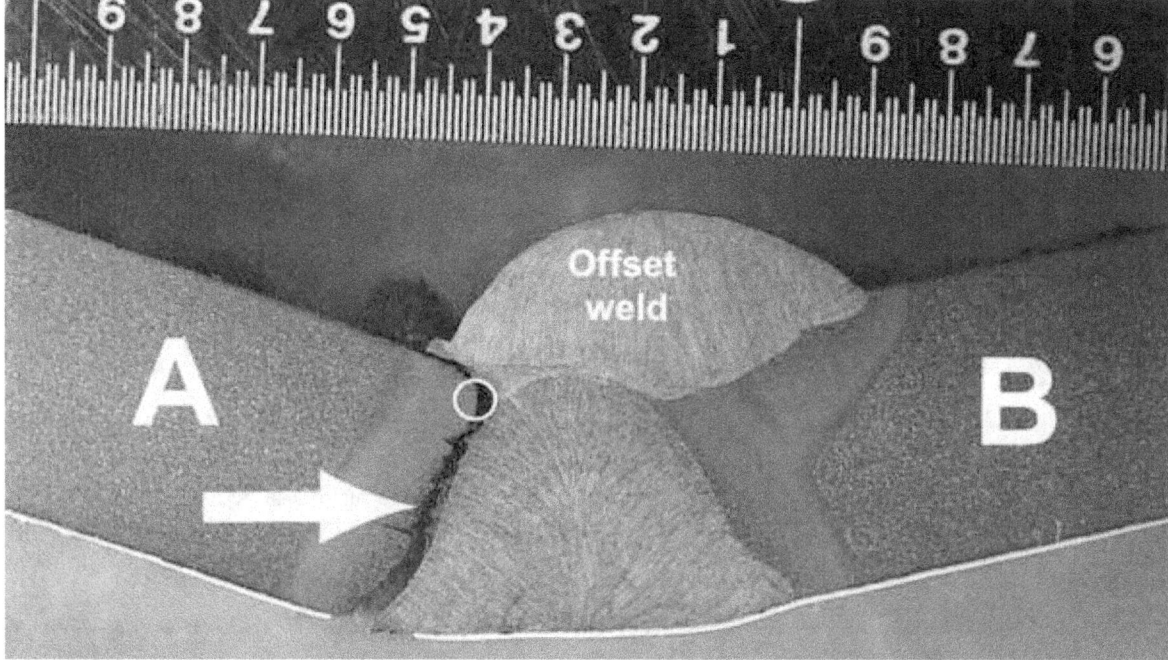

Figure 6. Cross section view of nurse tank longitudinal weld, showing left steel shell plate A and right steel shell plate B welded together. Fracture through tank shell is indicated by arrow. Circle marks area of lack of fusion.

Postaccident examination and matching of the characteristics on each side of the longitudinal fracture in the shell of the accident nurse tank revealed that the two sides of the shell could not be aligned to form a circular arc. Where they had been welded together, the two sides were at about a 30-degree angle relative to each other.

Other Postaccident Examinations and Testing

After the accident, the nurse tank's pressure relief device was found on the tank with a dust cover still in place over the device opening.[13] The device was tested at an independent laboratory using argon gas. The device activated between 260 and 261 psig, within the required tolerances.

The nurse tank's float gauge was examined at the manufacturer's facility under Safety Board supervision. The float gauge dimensions were compared to, and determined to meet, the manufacturer's design specifications. The moving parts (gears, shafts, magnet, and indicator needle) were examined and tested; they were operating within design specifications.

The nurse tank's dip tube was measured by the Iowa Department of Agriculture and Land Stewardship and found to have a length of 9 3/4 inches. (According to Trinity drawings, the standard dip tube length was 9 1/2 inches with the tank 85 percent full.)

Anhydrous Ammonia

According to the material safety data sheet used by CF Industries, Inc., which was River Valley's anhydrous ammonia supplier, anhydrous ammonia (NH_3) is a poisonous and corrosive gas with a boiling temperature of -28° F. Anhydrous ammonia is transported as a liquefied compressed gas in pressure vessels (tank cars, cargo tanks, nurse tanks, and cylinders). Pressurization keeps the anhydrous ammonia liquid while in the pressure vessel. If it is released from the pressure vessel, anhydrous ammonia will immediately return to a gaseous state and expand rapidly. At 80° F, about the reported temperature at the time of the Calamus accident, the vapor pressure of anhydrous ammonia would result in an internal tank pressure of 142 psig.[14]

Under DOT regulations (49 *Code of Federal Regulations* [CFR] Parts 171–180), anhydrous ammonia is classified and regulated for domestic shipment as a nonflammable gas but is required to be identified as an "inhalation hazard." For international shipment,

---

[13] A *dust cover* is a soft rubber cap that fits over the opening of the pressure relief device to keep out dust, debris, and water. It is held in place by friction. The dust cover is designed to blow off the opening if internal tank pressure causes the pressure relief device to activate.

[14] The *vapor pressure* of a liquefied compressed gas is the pressure exerted by vapors in equilibrium over the liquefied form in a closed container. Vapor pressure thereby provides a measure of internal tank pressure when the liquefied gas is at a given temperature.

the DOT classifies anhydrous ammonia as a poisonous gas, with a subsidiary hazard classification of class 8, corrosive.

According to *Medical Management Guidelines for Ammonia*, issued by the U.S. Department of Health and Human Services' Agency for Toxic Substances and Disease Registry,[15] anhydrous ammonia has the following health effects:

> Ammonia is highly irritating to the eyes and respiratory tract. Swelling and narrowing of the throat and bronchi, coughing, and an accumulation of fluid in the lungs can occur.

> Ammonia causes rapid onset of a burning sensation in the eyes, nose, and throat, accompanied by lacrimation [discharge of tears], rhinorrhea [runny nose], and coughing. Upper airway swelling and pulmonary edema may lead to airway obstruction.

> Prolonged skin contact (more than a few minutes) can cause pain and corrosive injury.

According to the National Institute for Occupational Safety and Health (NIOSH),[16] the "low lethal" concentration (LC_{LO})[17] of anhydrous ammonia for humans is 5,000 parts per million (ppm) for a period of 5 minutes. NIOSH also stipulates that the "immediately dangerous to life or health" (IDLH) concentration[18] of anhydrous ammonia is 300 ppm. Humans can detect the odor of anhydrous ammonia at 3 to 5 ppm.

River Valley Cooperative

River Valley is a farmer-owned cooperative with about 1,800 members/owners. It maintains 11 active anhydrous ammonia filling facilities, all in eastern Iowa. It owns about 600 nurse tanks used in its anhydrous ammonia service. The cooperative provides its members propane, gasoline, diesel, livestock feed, crop nutrients, herbicides, and insecticides. In addition, the cooperative provides grain storage and conducts marketing for its members.

---

[15] See <http://www.atsdr.cdc.gov/MHMI/mmg126 html> for the complete medical management guidelines.

[16] *NIOSH* is a Federal agency responsible for conducting research and making recommendations for the prevention of work-related injury and illness. It is part of the Centers for Disease Control and Prevention within the U.S. Department of Health and Human Services.

[17] The LC_{LO} is the lowest concentration of a substance that has been reported to cause death in humans.

[18] In its hazardous waste operations and emergency response regulation (29 CFR 1910.120), the Occupational Safety and Health Administration defines an *IDLH concentration* as "An atmospheric concentration of any toxic, corrosive or asphyxiant substance that poses an immediate threat to life or would cause irreversible or delayed adverse health effects or would interfere with an individual's ability to escape from a dangerous atmosphere." NIOSH, in its "Respirator Decision Logic," defines *IDLH exposure condition* as a condition that poses a threat of exposure to airborne contaminants when that exposure is likely to cause death or immediate or delayed permanent adverse health effects or prevent escape from such an environment.

The River Valley Calamus facility was used solely for anhydrous ammonia filling service. Farmers using the Calamus facility for anhydrous ammonia service typically dropped off an empty nurse tank and picked up a full one.

Nurse Tank Loading Procedures

River Valley's Calamus facility manager indicated that loaders generally used uniform procedures when filling a tank with anhydrous ammonia at the loading platform. He said the loaders would:

- Weigh the empty nurse tank. (Using the tank scale beside the facility office.)

- Pull the tank to the loading platform using a pickup truck.

- Unhook the pickup truck and pull it away.

- Hook up the hoses. (The loaders were to attach the red fitting to the red valve for the liquid line and the yellow fitting to the yellow valve for the vapor return line.) (See figure 7.)

- Open both valves on the top of the nurse tank.

- Open both valves on the hose ends.

- Open both valves on the platform. (Loaders were told that any leak would cause an odor and, if they detected an odor, they should shut the valves down.)

- Watch the float gauge on top of the nurse tank and shut down the valves when the gauge reached 85 percent. (The nurse tank loader who survived the accident said that in addition to watching the float gauge, he routinely checked the level of anhydrous ammonia in each tank using the dip tube to ensure the tank was not overfilled.)

- Shut down the valves in reverse order. (Platform to hose to nurse tank.)

- Hook the full tank to the pickup truck and reweigh it to determine the amount of anhydrous ammonia that had been loaded into the tank.

Figure 7. A nurse tank hooked to loading hoses.

The facility manager stated that loaders typically loaded four nurse tanks simultaneously. When loading multiple tanks, loaders were to walk around and check the gauges.

The nurse tanks were placarded "Nonflammable Gas." No shipping papers were prepared, and none were required.

Nurse Tank Periodic Inspection

River Valley told investigators that it carried out external visual inspections of its nurse tanks, including the accident tank, annually before each spring planting season. There is no record that River Valley had conducted any internal inspections, periodic or random, on the nurse tank since its manufacture. There is no Federal or State requirement that periodic inspections be conducted on nurse tanks. (See the "Regulations and Standards Affecting Nurse Tanks" section of this report for additional information on inspection requirements.)

Emergency Procedures and Equipment

River Valley. The River Valley Calamus facility had no written procedures detailing what loaders should do in case of a release of anhydrous ammonia. The facility manager said that he told facility loaders that if an anhydrous ammonia release occurred, they were to hit an emergency shutoff button on the loading platform (which turns off the anhydrous ammonia pump) and get into the water immersion tub. (See figure 8.) Loaders were to remain in the tub until emergency responders arrived and told them it was safe to get out. Water immersion tubs are designed to provide on-site emergency first aid for injuries caused by exposure to corrosive materials by enabling a person whose skin or soft tissues are exposed to the material to flush the area with water. They are not intended to provide protection from, or first aid for, inhalation injuries.

Figure 8. Immersion tub in the center of loading platform. Arrow indicates emergency shutoff button location.

The tub held 200 gallons of water and was large enough for two individuals to sit in it, but it would have been difficult for two occupants to fully immerse themselves. A second immersion tub was located at one end of the facility's storage tank, about 40 feet from the loading platform.

River Valley officials told Safety Board investigators that, as of May 2004, the organization had developed no new procedures for its employees to follow when facing a major release of anhydrous ammonia.

Other. The CF Industries' November 2002 material safety data sheet for anhydrous ammonia states that in response to inhalation and eye exposure to anhydrous ammonia, responders should "Immediately move victim away from exposure and into fresh air." It further states that when skin is exposed to anhydrous ammonia, responders should "Immediately flush affected area(s) with large amounts of water ...."

A NIOSH study entitled *HAZOP of Anhydrous Ammonia Use in Agriculture*[19] recommends that:

> Untrained personnel are not to attempt to stop a release in the event of a broken nurse tank hose. Untrained personnel are instead to: 1) Immediately vacate the area by heading upwind ....

According to the *2000 Emergency Response Guidebook*,[20] in case of a small spill of anhydrous ammonia "from a small package or a small leak from a large package," people should evacuate to a distance at least 100 feet (in all directions) from the spill/leak source. In case of a large spill (from a "large package or many small packages"), people should withdraw at least 200 feet from the source.

Regulations and Standards Affecting Nurse Tanks

The DOT and the U.S. Department of Labor's Occupational Safety and Health Administration (OSHA) both have regulations covering nurse tanks. Both sets of regulations reference the requirements of the ASME *Boiler and Pressure Vessel Code* as they apply to nurse tanks.[21] The ASME requirements are detailed later in this section.

U.S. Department of Transportation

The DOT's Research and Special Programs Administration (RSPA) promulgates the Hazardous Materials Regulations (49 CFR Parts 171–185).

Regulations Covering Initial Manufacture and Qualification of Cargo Tanks. Under the Hazardous Materials Regulations, the manufacturing specifications for DOT-specification cargo tanks used to transport anhydrous ammonia (MC 331 tanks) incorporate the requirements of the ASME *Boiler and Pressure Vessel Code*. Also, these

---

[19] Thomas McKelvey, *HAZOP of Anhydrous Ammonia Use in Agriculture*, prepared for NIOSH by Technica, Inc., Columbus, Ohio, May 1991.

[20] The *Emergency Response Guidebook* was developed jointly by the DOT, Transport Canada, and the Secretariat of Communications and Transportation of Mexico for use by firefighters, police, and other emergency services personnel who may be the first to arrive at the scene of a transportation incident involving a hazardous material. It is primarily a guide to assist first responders in (1) quickly identifying the specific or generic classification of the material(s) involved in the incident and (2) protecting themselves and the public during this initial response phase. The guidebook is updated every 3 to 4 years to accommodate new products and technology. The next update is scheduled for 2004.

[21] The OSHA regulations specifically cite Section VIII of the ASME *Boiler and Pressure Vessel Code*, entitled "Rules for Construction of Pressure Vessels," which concerns pressure vessels, including nurse tanks. The DOT regulations for nurse tanks do not specifically cite this section of the ASME code, but they reference Section VIII (and several other sections unrelated to nurse tank manufacture). DOT representatives told Safety Board investigators that the part of the ASME code applicable to the DOT regulations is Section VIII.

cargo tanks must meet numerous additional requirements established by the DOT,[22] including weld testing and inspection requirements. Under the weld requirements, the MC 331 tank welds must be subjected to one of a series of nondestructive inspections and tests to identify weld defects. The acceptable inspection and testing methods include wet fluorescent magnetic particle inspection, radiography, liquid dye penetrant testing, and ultrasonic testing.

The nonspecification cargo tanks known as nurse tanks must also be manufactured in accordance with the ASME *Boiler and Pressure Vessel Code*. However, as long as nurse tanks meet the general criteria set forth in section 173.315(m) of the Hazardous Materials Regulations,[23] they are not required to meet any additional DOT-specified requirements.

Regulations Covering Periodic Inspections. Under the DOT Hazardous Materials Regulations, all DOT-specification bulk containers for liquids and liquefied compressed gases that carry hazardous materials and meet DOT specifications—including railroad tank cars, cargo tanks, and intermodal bulk containers—must be periodically inspected and tested to ensure they are structurally sound. DOT-specification cargo tanks must adhere to a schedule of periodic inspections. DOT-specification MC 331 cargo tanks, which carry anhydrous ammonia, are inspected on the following schedule:

External visual inspection	Annual
Internal visual inspection	Every 5 years
Leakage test	Annual
Pressure test	Every 5 years

Until 1981, the DOT allowed nonspecification cargo tanks with the same configuration as nurse tanks to be manufactured for the transportation of liquefied petroleum gas.[24] In 1981, the regulations were amended to prohibit new manufacture of nonspecification tanks for liquefied petroleum gas service. Title 49 CFR 173.315(k) allowed the continued use of nonspecification cargo tanks in liquefied petroleum gas service in intrastate commerce. Section 173.315(k)(5) required that these tanks in liquefied petroleum gas service be periodically inspected and hydrostatically tested in

---

[22] Specification MC 331 is found at 49 CFR 178.337. This section addresses cargo tank motor vehicles used for transportation of compressed gases. It includes sections on materials specifications and on impact, pressure, and stress testing procedures.

[23] This section states that a nurse tank does not have to meet the requirements of DOT-specification cargo tanks if it: (1) Has a minimum design pressure of 250 psig and meets the requirements of the edition of the ASME code in effect at the time it was manufactured and is marked accordingly; (2) Is equipped with safety relief valves meeting the requirements of Compressed Gas Association pamphlet S1.2; (3) Is painted white or aluminum; (4) Has capacity of 3,000 gallons or less; (5) Is loaded to a filling density no greater than 56 percent; (6) Is securely mounted on a farm wagon; and (7) Is in conformance with the requirements of part 172 of this subchapter except that shipping papers are not required; and it need not be marked or placarded on one end if that end contains valves, fittings, regulators or gauges when those appurtenances prevent the markings and placard from being properly placed and visible.

[24] Propane is the liquefied petroleum gas most commonly transported in these tanks.

accordance with DOT specifications for MC 331 cargo tanks. However, the DOT Hazardous Materials Regulations do not require periodic inspection or testing of nonspecification cargo tanks used for anhydrous ammonia service (nurse tanks).

U.S. Department of Labor Occupational Safety and Health Administration

OSHA regulations (29 CFR Part 1910–*General Industry*) provide, as stated under section 1910.111, general standards for the construction of tanks used to hold anhydrous ammonia. Subsection 1910.111(b)(2) provides the requirements for the construction, original testing, and requalification of various types of nonrefrigerated containers, including farm vehicles used in the application of ammonia, such as nurse tanks. These standards set minimum manufacturing requirements and state that all such tanks shall be constructed and qualified in accordance with the ASME *Boiler and Pressure Vessel Code,* Section VIII, "Rules for Construction of Pressure Vessels."

Nothing in the OSHA regulations requires periodic inspection or testing of nurse tanks.

OSHA regulations also require that facilities containing stationary storage tanks for anhydrous ammonia, such as the large storage tank at the River Valley Calamus facility, have an easily available shower or 50-gallon drum of water.[25]

American Society of Mechanical Engineers

The ASME *Boiler and Pressure Vessel Code,* Section VIII, "Rules for Construction of Pressure Vessels," establishes manufacturing and qualification testing requirements for some types of pressure vessels, including nurse tanks. This part of the ASME code covers a wide range of vessels used in a variety of industrial applications, from large fixed storage vessels to portable compressed gas cylinders.

The ASME code provides a formula for design and fabrication of pressure tanks that includes such factors as tank wall thickness, tank service pressure, tank radius, stress values for steel, and joint efficiency of the tank welds. The joint efficiency is determined by the scope of the examination performed on the welds to ensure their quality. Manufacturers may choose to use full radiography, spot radiography, or no radiography to examine the welds.[26] Typically, one of two options—full radiography or spot radiography—is selected. Full radiography means that longitudinal welded joints must be radiographically examined along their full lengths. When conducting spot radiography, a 6-inch-long radiograph must be taken for every 50 feet of longitudinal weld. Spot radiography does not require a radiograph to be taken on the longitudinal weld of each

---

[25] Subsection 1910.111(b)(10)(iii).

[26] Technically, under the ASME code, a manufacturer may choose not to conduct any type of radiographic test of the weld, provided the other elements in the manufacturing formula (tank thickness, etc.) are so adjusted as to compensate for the lack of weld examination. According to Trinity, in practical terms, if a manufacturer did not use either full or spot radiography to ensure joint efficiency, it would not be economically feasible for the manufacturer to produce a competitive tank.

nurse tank manufactured. (On nurse tanks manufactured by Trinity, the longitudinal welds typically range in length from 12 to 15 feet.)

Under the ASME code, if manufacturers choose to use spot radiography instead of full radiography to ensure the quality of tank welds, their tank design strength calculations are affected in such a way that for a given design pressure, tanks must either be constructed of thicker materials, constructed of higher strength steels, or constructed with some combination of both thicker and higher strength steels to compensate for the less rigorous weld inspection provided by spot radiography.

The ASME code requires that full radiography be used to qualify the welds on tanks intended to contain "lethal substances." ASME defines lethal substances as "poisonous gases or liquids of such a nature that a very small amount of the gas or of the vapor of the liquid mixed or unmixed with air is dangerous to life when inhaled." ASME does not identify any specific materials as lethal substances. The tank user is required to tell the designer or manufacturer of the tank if the tank is to be used to hold a lethal substance. Historically, tank users have not identified anhydrous ammonia as a lethal substance.

The ASME code states that any weld containing an area of incomplete fusion or penetration is unacceptable. It also states that other indications[27] shall not exceed two-thirds the thickness of the weld, with a maximum acceptable length of 3/4 inch.

The ASME code also limits the permissible out-of-roundness of a tank shell to 1 percent of the nominal diameter.

The ASME code in effect at the time of the accident nurse tank's manufacture required, as the final step in the manufacturing qualification process, that each tank be hydrostatically pressure-tested after all fabrication was completed and all examinations were performed.[28] The required test pressure was 1.5 times the maximum allowable working pressure (375-psig test pressure for a 250-psig tank).[29]

Neither the ASME code in effect at the time of the accident nurse tank's manufacture nor the current code requires periodic inspection or testing of pressure vessels, such as the nurse tank.

---

[27] The term "other indications" refers to other areas of a weld identified by the radiograph as having problems or flaws.

[28] Including radiographic examinations of the welds.

[29] The current edition of the ASME code requires each newly manufactured tank to be hydrostatically pressure-tested to 1.3 times the maximum allowable working pressure (325-psig test pressure for a 250-psig tank).

Other Information

Nurse Tanks in the United States

General. A director of Government Relations for The Fertilizer Institute told investigators that the organization estimates that about 200,000 nurse tanks are currently in use in the United States. The institute also estimates that about one to five nurse tanks are retired annually and considers it probable that some tanks that were manufactured in the 1950s and 1960s remain in service today. The institute estimates that the tanks in service deliver about 3.9 million short tons (in the range of 1.0 million to 1.5 million nurse tank loads) of anhydrous ammonia to farmers' fields annually.

Nurse tanks have no established regulatory limit to their service lives. Provided the tanks continue to perform satisfactorily and have no obvious external defects, the tanks likely will remain in service.

RSPA maintains data on nurse tank accidents occurring in transportation. Upon reviewing this data, investigators found no recorded instance of a nurse tank accident resulting from a longitudinal weld or other structural weld failure. Most nurse tank accidents recorded by RSPA involved hose breaks, component failures, and rollovers. However, nurse tanks are used in farm environments, and accidents may have taken place when the tanks were not in transportation (that is, being transported over public roads). Such accidents might not have been reported to RSPA. For example, the Calamus accident, which took place at a filling facility, did not appear in RSPA's accident database.

The Fertilizer Institute does not collect information on nurse tank accidents, and nurse tank manufacturers could provide no reliable information to investigators regarding nurse tank accident rates. Neither Iowa nor Minnesota (both significant farming States) could provide data on nurse tank accidents in their States beyond some anecdotal information.

Minnesota Department of Agriculture Information. In September 1996, the Minnesota Department of Agriculture began a program to address a statewide problem of illegible or missing nameplates on nurse tanks used for handling anhydrous ammonia. The State allowed the tanks to be certified and re-nameplated, provided they passed certain nondestructive tests. The testing for each nurse tank, performed by ASME-certified pressure vessel repair shops, had three successive stages: first, a thorough visual examination to detect obvious defects in the tank; then, metal thickness testing on the heads and shell to determine that the calculated maximum allowable working pressure (MAWP) equaled or exceeded the Federal MAWP of 250 psig; and, finally, a hydrostatic pressure test to 375 psig. As each nurse tank passed a test stage, it moved on to the next. If a tank failed a given test stage, the State permitted it to undergo ASME-authorized repairs to correct problems, after which the tank was retested. If a tank could not be repaired, it was removed from service.

According to State officials, between 1,500 and 2,000 nurse tanks were tested under this program. State officials told investigators that some of the tanks tested did not pass the initial testing. The primary defects found with the tanks included external corrosion, tank dents, and small leaks. The State did not record the number of nurse tanks that failed the initial testing and were repaired, nor did the State require the repair shops to forward reports on nurse tanks that were removed from service as a result of the retesting. However, the State located records showing that at least 10 nurse tanks failed the retesting and were removed from service.

Of these 10 tanks removed from service, 3 were removed as a result of defects noted during the visual examination: 1 had a severe dent in the rear head; 1 had a bulge in the rear head; and 1 had an attachment welded to the head that was not part of the tank's original design. Six tanks were removed from service because the metal thickness in the shell, heads, or both was too thin, and the tank could not be certified to a MAWP of 250 psig. One tank, having passed the first two stages of testing, failed the hydrostatic pressure test when an unrepairable leak was discovered in the rear head.

In addition, Minnesota State officials told investigators that in 1995 two nurse tanks in the State had catastrophic failures. The two tanks had been manufactured by Trinity in 1973 and had proximate serial numbers (832660 and 832656), indicating that they likely were produced in the same manufacturing lot. State officials said the running gear on the two nurse tanks was unusual in that the wheel sets were mounted directly to each of the tank ends, without any cross support between them. Both tanks split open at the middle circumferential weld, one in a storage yard and one as it was being transported on the road. Although no failure analysis was performed on these tanks, the State considered that a manufacturing defect and/or the unusual design (which may have allowed the middle of the tank to bend under load) may have been factors in their failures. RSPA's database contained no record of either of these two accidents.

New Trinity Manufacturing Procedures

Trinity had no record of the specific quality control procedures its Beardstown plant used to ensure tank roundness in 1976 when the accident nurse tank was manufactured. However, at that time, Trinity's general policy was to test each nurse tank for roundness after welding. A rod appropriately sized for the tank's internal diameter was placed (held horizontally) inside each tank following welding. The rod was rotated and, if it did not rotate freely, Trinity knew that an out-of-roundness condition existed. Trinity then conducted the necessary calculations to determine whether the degree of the tank's out-of-roundness was greater than the ASME-accepted level of 1 percent of the tank's nominal diameter. If it was, the tank was rejected. Trinity told investigators that it discarded this testing procedure some years ago because it was not finding any incidence of tanks being rejected for out-of-roundness.

For some years, Trinity has been requiring that its plants check tank roundness with a roundness template before the tank's longitudinal seam is welded. The procedure is used to ensure that each tank will meet the ASME standard limiting the out-of-roundness of a tank shell to 1 percent of the nominal diameter.

Full Examination of Nurse Tank Welds

Information from the two manufacturers of nurse tanks indicates that in the mid-1980s, a technology known as "radioscopy," which permits direct observation of objects opaque to light by means of some other form of radiant energy, such as x rays, was developed for use in the industry. This technology allows real-time examination of the entire longitudinal weld on nurse tanks. Radioscopy is, by definition, a radiographic method of examining welds. The industry is using radioscopy to conduct full radiographic examinations of the longitudinal welds on nurse tanks.

Adopting radioscopy as the means of checking the full longitudinal weld enabled manufacturers to use thinner materials in the manufacture of nurse tanks. This provided an economic incentive for the use of radioscopy in lieu of spot radiography. According to the two companies, all nurse tanks manufactured since the mid-1980s have received 100-percent radioscopic examination of their longitudinal welds at the time of manufacture.

Analysis

This analysis is presented in three main parts. First, the Safety Board identifies factors that can be readily excluded as causal or contributory to the accident. In the second part, the Board analyzes the causes and factors contributing to the accident. In the remainder of the analysis, the Board addresses the safety issues arising from the investigation, which are as follows:

- The adequacy of standards for initial qualification and periodic testing of nurse tanks.

- The adequacy of River Valley's emergency procedures for anhydrous ammonia nurse tank loaders.

Exclusions

The Safety Board considered whether the nurse tank might have been overpressurized on the day of the accident. Although investigators were unable to determine the exact procedures the loaders used to fill the nurse tank involved in the accident, the surviving loader stated that he typically used the dip tubes installed on the tanks to ensure that the tanks were not overfilled. The dip tube on the accident nurse tank was examined after the accident and determined to be an appropriate size for the tank. The nurse tank had been filled and disconnected from the filling apparatus without any difficulty before the tank failure occurred.

Further, the dust cover was found still covering the opening of the nurse tank's pressure relief device following the accident. If the device had activated due to excess pressure within the tank, the dust cover would have been blown off the device. The device was tested at an independent laboratory and found to operate within specification. Therefore, the Safety Board concludes that the nurse tank was not overpressurized.

The effectiveness of the emergency response was also examined. Representatives of the local emergency response agencies were on site and providing assistance within about 10 minutes of their notification. Medical evacuation helicopters arrived at the accident site soon after being called and quickly transported the injured to hospitals.

The Accident

This accident was caused by a fracture in the shell of the nurse tank at the longitudinal weld, which resulted in a complete release of the anhydrous ammonia from the tank.

Postaccident metallurgical examination indicated that a portion of the nurse tank's interior longitudinal weld was not centered on the shell seam. Further, in one area where the inner weld was offset to one side of the shell seam, there was a 3.25-inch-long and up to 0.102-inch-deep region of incomplete fusion. This area would have increased the local stresses experienced by the tank shell when the tank was pressurized.

In addition, the failed section of the nurse tank had some indications of an out-of-roundness condition. Postaccident examination indicated that, during the nurse tank's manufacture, the two sides of its shell that formed the longitudinal seam may have been improperly aligned. When examined after the accident, the two sides could not be aligned to form a circular arc and appeared to be aligned at about a 30-degree angle relative to each other at the weld. This indicates that there was some misalignment of the shell surfaces when the tank was manufactured. A misalignment would have produced an area of localized stress concentration and increased tensile stress on the inner surface of the shell when the tank was pressurized. The Safety Board concludes that the unfused area and offset portion of the longitudinal weld significantly weakened the nurse tank shell; misalignment of the shell surfaces and the large angle weld bead increased the operating stresses in the weakened area.

Examination of the area of black oxidation below the unfused area of the weld showed that a crack initiated from the unfused area, moving from the interior surface of the tank toward the outside surface, as indicated by its curved shape encompassing the unfused area. This 12.75-inch-long oxidized area, in combination with the unfused area, penetrated 83 percent of the tank shell, leaving only about 0.063 inch of steel to maintain the tank's integrity. Because the process of oxidation in a low-oxygen environment takes place over time, the black oxidation covering the fracture surface of this part of the crack indicates that the crack was present for some time before the final fracture occurred on April 15, 2003. Also, the relatively rough surface of the bulk of the oxidized area[30] means this part of the crack likely formed as a result either of a single overstress event or of low-cycle fatigue.

The proof pressure test that took place at the conclusion of the 1976 manufacture of the accident nurse tank would have provided an elevated stress event. Given the identified weaknesses of the tank shell in the region containing the unfused area and the offset weld, the stress caused by the 375-psig proof pressure test would likely have been sufficient to initiate the crack. It is also likely that some portion of the fracture surface found covered with black oxidation had opened during the initial proof pressure test and had been within the tank shell since the tank's manufacture in 1976.

Although the proof pressure test probably initiated the rough-featured part of the oxidized portion of the crack, this test may not have been the only event that contributed to its extent. Some of this portion of the crack may have been formed by low-cycle fatigue events occurring during the service life of the tank, and the evidence of that fatigue may subsequently have been obliterated by corrosion. The Safety Board was unable to develop

---

[30] A 0.007-inch-wide region on the outer edge of the oxidized area showed evidence of high-cycle fatigue.

a full and accurate history of this nurse tank, but it is possible that some low-cycle fatigue event, such as a high-force impact accident or an additional proof pressure test, could have further extended the blackened, oxidized portion of the crack from its extent following the 1976 proof pressure test to its eventual size. However, an accident of sufficient magnitude to extend the crack would likely have left physical evidence of damage and repair on the tank, and none was found. Also, proof pressure tests are rarely conducted on nurse tanks after the initial proof pressure test, and no record was found indicating that such a test was performed on this tank after its manufacture in 1976.

The crack continued to grow by fatigue as the tank continued in service over 27 years, being repeatedly loaded to about 140 psig[31] and unloaded. This process is evidenced by the relatively smooth fracture features with curving boundaries typical of high-cycle fatigue in the 0.007-inch-wide region on the outer edge of the oxidized area of the crack.

Then, on April 15, 2003, the nurse tank failed when the crack became unstable and propagated through the remaining thickness of the shell under the pressure from the routine loading of the tank with anhydrous ammonia on that day. This is evidenced by the fact that the final region in the fracture shows indications of overstress failure.

Based on the foregoing analysis of the postaccident metallurgical findings, the Safety Board concludes that a crack initiated from the unfused area of the longitudinal weld in the nurse tank, most likely during the manufacturing proof pressure test, and grew through the tank shell by fatigue until the tank shell failed under normal operating conditions.

As a result of these findings, the Safety Board examined the adequacy of regulations and standards for the manufacturing and testing of nurse tanks used for the transportation of anhydrous ammonia. The Safety Board also examined River Valley's procedures for responding to an anhydrous ammonia release.

Standards for Manufacturing and Testing Nurse Tanks

Initial Qualification of Welds

Trinity stated that the accident nurse tank was manufactured in 1976, in adherence with the standards in the ASME *Boiler and Pressure Vessel Code*, Section VIII. As permitted by the code, Trinity used spot radiography to qualify the longitudinal welds on its nurse tanks. The use of spot radiography meant that only one representative 6-inch-long radiograph was used to ensure the quality of every 50 feet of Trinity's nurse tanks' longitudinal welds. In other words, this method tested only 1 percent of the length of the longitudinal welds on Trinity's nurse tanks. Further, because the code does not specify that a spot radiograph must be taken from each nurse tank and because Trinity's

---

[31] The loading pressure would have been about 140 psig on a day such as the one on which the accident occurred, with the temperature around 80° F. Somewhat lower pressures would have been experienced on days with lower temperatures, and somewhat higher pressures would have occurred on warmer days.

tank shell lengths range from 12 to 15 feet, a single 6-inch-long radiograph could have been used to qualify the longitudinal welds on three or four nurse tanks. Consequently, most of the tanks produced by Trinity during this period received no individual radiographic testing of their welds.

Use of the spot radiography method to qualify the longitudinal welds on nurse tanks allows the majority of these welds to go uninspected and significantly reduces the likelihood that critical defects and flaws that would result in the rejection of the welds will be detected. Had a full radiographic examination been made on the accident nurse tank during its manufacture, the 3.25-inch-long weld flaw it contained likely would have been detected and the tank repaired or rejected. The Safety Board concludes that using spot radiography to qualify longitudinal welds in nurse tanks manufactured to transport anhydrous ammonia, a poisonous and corrosive gas, is not a sufficiently reliable method of detecting critical flaws that can result in tank failure.

Since the mid-1980s, as the result of the development of radioscopy, a radiographic technology that allows real-time examination of the longitudinal weld, the two manufacturers of nurse tanks have chosen to examine the full lengths of the longitudinal welds on all their nurse tanks using radioscopy. Thus, this full radiography method of qualifying longitudinal welds has become general practice in the nurse tank manufacturing industry. However, under existing regulations, the manufacturers of nurse tanks can, if they choose, return to using spot radiography to verify longitudinal tank welds. Although it does not appear likely that nurse tank manufacturers will return to using spot radiography to qualify longitudinal welds, if they chose to do so it would increase the risk that they would fail to identify weld defects during the manufacturing qualification process. Given the serious consequences of a major anhydrous ammonia release, which could be caused by a weld defect, the Safety Board urges RSPA to monitor nurse tank manufacturers to ensure that they continue to use a full radiography method of qualifying the longitudinal welds of their nurse tanks.

Periodic Testing

After being manufactured and proof pressure-tested, the accident nurse tank was never required to undergo periodic testing of any kind to ensure its safety during its service life. As was noted in a previous section of this analysis, the initial crack in the nurse tank shell was likely introduced during the nurse tank's original proof pressure test in the manufacturing process in 1976.[32] Because the crack was not visible on the outside surface of the tank, exterior examination during or after manufacture would not have detected the crack. Because the nurse tank had no openings larger than about 2 inches in diameter and internal baffles blocked portions of the interior shell from visual examination through those openings, attempts at normal internal visual inspection would also have been unsuccessful. However, given the crack's considerable size and the fact that it penetrated 83 percent of the tank shell thickness, it could have been detected during the

---

[32] Because ASME requires proof pressure testing to be performed after all manufacturing and testing has been completed, the crack did not exist when the initial manufacture spot radiography qualification of the accident nurse tank's welds took place.

service life of the tank by a variety of nondestructive testing methods. The Safety Board concludes that periodic nondestructive testing could have detected the weld defect and internal crack in the nurse tank during its service life, and the tank could have been repaired or removed from service before it failed.

Neither the DOT, OSHA, nor ASME has requirements for periodic inspection and testing of nurse tanks. Although the DOT Hazardous Materials Regulations establish periodic testing requirements for specification cargo tanks and all other specification bulk containers used to transport hazardous materials (including anhydrous ammonia), nurse tanks are excepted from these requirements. In fact, with the single exception of nurse tanks, all nonspecification cargo tanks built to the same configuration as nurse tanks are required to have periodic inspection and testing.

Of bulk containers that are used to transport hazardous materials, only nurse tanks are allowed to transport anhydrous ammonia without being required by the DOT to undergo some type of periodic inspection and/or testing during their service lives to ensure tank integrity. Nurse tanks, like any other cargo tank in pressurized service, experience stress and wear as they undergo repeated pressure cycles over months and years of service. They also are often transported over back roads, where pavement surfaces may be rough, as well as into fields. Such transport environments could put additional stress on the tanks. Some deterioration of nurse tank condition after years of service under these conditions seems inevitable.

According to The Fertilizer Institute, an estimated 200,000 nurse tanks are in service today. These apply between 1.0 million and 1.5 million loads of anhydrous ammonia to fields annually. The institute also estimates that only one to five nurse tanks are removed from service each year. Because they are not required to be removed from service at a given age, many nurse tanks that have received no effective safety inspections for several decades likely remain in use on farms and at filling facilities. In addition, based on the practices used by Trinity and other nurse tank manufacturers, it appears that many of the tanks that were manufactured before the mid-1980s did not receive full radiographic examination of their longitudinal welds upon manufacture. As shown by this accident, an undetected flaw in a longitudinal weld can cause a serious accident many years after a nurse tank's manufacture.

Further, the information provided by the Minnesota Department of Agriculture's unique 1996 program indicates that some nurse tanks in use may have corrosion, denting, or leaking problems requiring repair. Unless they are detected by an inspection and testing program, however, these defects might go unnoticed and continue to grow until a tank failure and anhydrous ammonia release occurs. The Minnesota program information also indicates that some of the defects found in the State's nurse tanks in the course of the 1996 program were so extensive the tanks could not be successfully repaired. At least 10 of the tanks tested eventually had to be removed from service because they could not be made sufficiently sound. Had Minnesota not detected these defective tanks through its own program, these tanks—containing serious defects—might still be in service today. Nurse tanks in other States may have integrity problems similar to those discovered in the Minnesota program.

Although failures of nurse tanks may be rare, when they occur, as in the case of the Calamus accident, they can be catastrophic, given the extremely hazardous nature of the anhydrous ammonia they contain. Therefore, the Safety Board believes that RSPA should require periodic nondestructive testing to be conducted on nurse tanks to identify material flaws that could develop and grow during a tank's service and result in a tank failure.

Tank Shell Out-of-Roundness

The ASME *Boiler and Pressure Vessel Code* limits the permissible out-of-roundness of a tank shell to 1 percent of the nominal diameter. Although the original geometry and roundness of the tank after welding is unknown, the Safety Board is concerned that, given the findings of the postaccident examination, the tank could have had a geometry with a significant out-of-roundness. Trinity is now using a roundness template on all its tank shells to ensure that each tank is not misaligned or out-of-round before the tank's longitudinal seam is welded. Trinity's current procedures require that any tank found to be out-of-round be sent back, reformed, and then checked again before that section of the tank is welded.

River Valley Emergency Preparedness

The sudden and complete release of anhydrous ammonia from a single nurse tank at the River Valley Calamus facility affected the safety only of the employees in the immediate vicinity of the accident—the two loaders. No one else was present, and the facility was in a remote location.

When the nurse tank split open at the Calamus facility, it quickly lost the bulk of its liquid contents. The spilled anhydrous ammonia rapidly vaporized, and for some minutes the vapor cloud probably enveloped the platform, the immersion tub, and the two loaders. When an anhydrous ammonia release occurs, it puts those in the vicinity at risk of two types of exposure—inhalation and skin/soft tissue exposure. The loaders evidently tried to follow the procedure that River Valley had told them to use in case of an anhydrous ammonia release—they tried to immerse themselves in the nearest water-filled immersion tub, which was on the loading platform. Immersion tubs enable workers to wash corrosive liquid from their skin and soft tissues. They are not designed to provide protection from or treatment for inhalation of chemical vapor. Although the loader immersed in the tub received some protection from skin/tissue exposure, both loaders were unable to avoid inhaling the ammonia vapor. The loader who was not initially immersed in the tub (and who subsequently died from his injuries) remained at risk not only to the inhalation hazards but also to the damaging effect of the gas on his skin.

Hazardous materials authorities have indicated that evacuation is the appropriate response to minimize inhalation exposure from significant anhydrous ammonia releases. The material safety data sheet for anhydrous ammonia provided by River Valley's anhydrous ammonia supplier states that when an inhalation exposure to anhydrous ammonia occurs, the victim should immediately be moved away from the exposure site and into fresh air. According to NIOSH, in its study *HAZOP of Anhydrous Ammonia Use in Agriculture*, workers should "Immediately vacate the area by heading upwind" when an anhydrous ammonia release occurs. The *2000 Emergency Response Guidebook* recommends that when a large spill of anhydrous ammonia takes place, people should move at least 200 feet away from the source. By contrast, the only procedure that River Valley told its loading employees to follow in the event of an anhydrous ammonia release directed immersion in the nearest water-filled tub, which was in the immediate proximity of the release from the nurse tank, leaving the loaders at risk of inhalation exposure. This was contrary to the guidance in the material safety data sheet, the NIOSH study, and the *2000 Emergency Response Guidebook*. On this basis, River Valley's instruction was ineffective as a response to a significant release such as can result from the failure of either a nurse or storage tank. Instead, River Valley should have directed its loaders, when faced with such a significant release, to evacuate the release area before taking steps to flush affected skin and tissue with water.

The Safety Board concludes that River Valley's emergency procedures were ineffective because they did not direct the nurse tank loaders to evacuate the area when an anhydrous ammonia release posed an inhalation hazard. To date, River Valley has not developed new procedures for its employees to follow in the event of a significant anhydrous ammonia release. Therefore, the Safety Board believes that River Valley should review manufacturers' material safety data sheets for anhydrous ammonia, NIOSH's *HAZOP of Anhydrous Ammonia Use in Agriculture*, and the *Emergency Response Guidebook* and establish written emergency procedures for employees to follow when an anhydrous ammonia release poses an inhalation hazard.

Conclusions

Findings

1. The nurse tank was not overpressurized.

2. The unfused area and offset portion of the longitudinal weld significantly weakened the nurse tank shell; misalignment of the shell surfaces and the large angle weld bead increased the operating stresses in the weakened area.

3. A crack initiated from the unfused area of the longitudinal weld in the nurse tank, most likely during the manufacturing proof pressure test, and grew through the tank shell by fatigue until the tank shell failed under normal operating conditions.

4. Using spot radiography to qualify longitudinal welds in nurse tanks manufactured to transport anhydrous ammonia, a poisonous and corrosive gas, is not a sufficiently reliable method of detecting critical flaws that can result in tank failure.

5. Periodic nondestructive testing could have detected the weld defect and internal crack in the nurse tank during its service life, and the tank could have been repaired or removed from service before it failed.

6. River Valley Cooperative's emergency procedures were ineffective because they did not direct the nurse tank loaders to evacuate the area when an anhydrous ammonia release posed an inhalation hazard.

Probable Cause

The National Transportation Safety Board determines that the probable cause of the sudden failure of the nurse tank at the anhydrous ammonia filling facility near Calamus, Iowa, on April 15, 2003, was inadequate welding and insufficient radiographic inspection during the tank's manufacture and lack of periodic testing during its service life.

Recommendations

As a result of its investigation of the April 15, 2003, hazardous materials accident near Calamus, Iowa, the National Transportation Safety Board makes the following safety recommendations:

To the Research and Special Programs Administration:

Require periodic nondestructive testing to be conducted on nurse tanks to identify material flaws that could develop and grow during a tank's service and result in a tank failure. (H-04-23)

To the River Valley Cooperative:

Review manufacturers' material safety data sheets for anhydrous ammonia, the National Institute for Occupational Safety and Health's *HAZOP of Anhydrous Ammonia Use in Agriculture*, and the *Emergency Response Guidebook* and establish written emergency procedures for employees to follow when an anhydrous ammonia release poses an inhalation hazard. (H-04-24)

BY THE NATIONAL TRANSPORTATION SAFETY BOARD

ELLEN ENGLEMAN CONNERS
Chairman

MARK V. ROSENKER
Vice Chairman

CAROL J. CARMODY
Member

RICHARD F. HEALING
Member

DEBORAH A. P. HERSMAN
Member

Adopted: June 22, 2004

this page intentionally left blank

Appendix A

Investigation

The National Transportation Safety Board's Communications Center intercepted an Internet news story about the accident about 11:30 a.m. eastern daylight time on April 17, 2003. An investigator was dispatched from Washington, D.C., to Calamus, Iowa. No Board Member went with the investigator. Investigative groups were established for hazardous materials and metallurgy.

Parties to the investigation were the Federal Motor Carrier Safety Administration; the Research and Special Programs Administration; the River Valley Cooperative; the Iowa Department of Agriculture and Land Stewardship; Trinity Industries, Inc.; and Continental NH_3 Products.

The Safety Board did not conduct a public hearing during this investigation.

this page intentionally left blank

www.ingramcontent.com/pod-product-compliance
Lightning Source LLC
Chambersburg PA
CBHW081805170526

45167CB00008B/3335